BREITLING®
HIGHLIGHTS

HENNING MÜTZLITZ

Schiffer Publishing Ltd®

4880 Lower Valley Road • Atglen, PA 19310

Other Schiffer Books by the Author:
Omega Highlights. ISBN: 978-0-7643-4212-7. $29.99

Other Schiffer Books on Related Subjects:
Breitling: The History of a Great Brand of Watches 1884 to the Present. Benno Richter. ISBN: 9780764326707. $49.95
Omega Designs: Feast for the Eyes. Anton Kreuzer. ISBN: 9780764329951. $59.99
Rolex Wristwatches: An Unauthorized History. James M. Dowling & Jeffrey P. Hess. ISBN: 0764324373. $125.00
Swiss Wristwatches: Chronology of Worldwide Success. Gisbert L. Brunner & Christian Pfeiffer-Belli. ISBN: 0887403018. $69.95
Vintage Rolex® Sports Models - 3rd Edition. Martin Skeet & Nick Urul. ISBN: 9780764329814. $79.99
Wristwatch Chronometers. Fritz von Osterhausen. ISBN: 0764303759. $79.95

Originally published by Heel Verlag GmbH as *Breitling Highlights* in 2009
Editor: Henning Mützlitz
English Translation: Elizabeth Doerr
Design and Layout: Muser Medien GmbH, Tanja Küppershaus

Copyright © 2012 by Schiffer Publishing, Ltd.

Library of Congress Control Number: 2012938699

Cover by Justin Watkinson
Type set in Tall Films Expanded/ITC Avant Garde Gothic Std

ISBN: 978-0-7643-4211-0
Printed in China

Schiffer Books are available at special discounts for bulk purchases for sales promotions or premiums. Special editions, including personalized covers, corporate imprints, and excerpts can be created in large quantities for special needs. For more information contact the publisher:

Published by Schiffer Publishing, Ltd.
4880 Lower Valley Road
Atglen, PA 19310
Phone: (610) 593-1777; Fax: (610) 593-2002
E-mail: Info@schifferbooks.com

For the largest selection of fine reference books on this and related subjects, please visit our website at **www.schifferbooks.com**
We are always looking for people to write books on new and related subjects.
If you have an idea for a book, please contact us at proposals@schifferbooks.com

This book may be purchased from the publisher.
Please try your bookstore first.
You may write for a free catalog.

In Europe, Schiffer books are distributed by
Bushwood Books
6 Marksbury Ave.
Kew Gardens
Surrey TW9 4JF England
Phone: 44 (0) 20 8392 8585
Fax: 44 (0) 20 8392 9876
E-mail: info@bushwoodbooks.co.uk
Website: www.bushwoodbooks.co.uk

BREITLING

CONTENTS

1................................ **Brand History**..................................Page 06
125 eventful years have made Breitling a globally reputable brand

2................................ **Historical Models**...........................Page 14
A selection of the most beautiful wristwatches of Breitling's brand history

3................................ **Navitimer**...Page 26
Breitling's most successful model has been a true companion for many pilots all over the world since 1952

4................................ **Avenger**..Page 40
Strong and robust: the Avenger unites masculine attributes and high precision

5................................ **Chronographs**...................................Page 50
Breitling offers chronographs for almost every need

6................................ **Superocean**......................................Page 64
With a Breitling on your wrist, you can conquer not only the skies, but also the depths of the seas

7................................ **"Breitling for Bentley"**..................Page 74
When precision and luxury meet, something really special is created

8................................ **Cockpit**...Page 86
Seeing it all at a glance: the Cockpit restricts itself to what matters

BREITLING
PREFACE

Breitling today is chiefly a symbol of masculinity and self-confidence, but also of quality, reliability, and — last, but certainly not least — precision. There is hardly another watch brand that has understood its history so well, credibly maintaining its identity, progress, and own demands upon itself thanks to the clever watchmakers at home in Grenchen. Breitling is not a generalist looking to satisfy every taste and need. Rather, Breitling concentrates on that which has set it apart from all others: making top instruments for a demanding clientele.

This book is designed to provide a view of that which sets —and has set — Breitling apart from the others. Alongside a short history of the brand, we present you with one hundred of the most remarkable models of the past and present. From the first monopusher chronograph to the legendary first editions of the Navitimer and the groundbreaking manufacture chronograph Chronomat B01 presented in 2009, you will find here exactly the models upon which the fame of the legendary Swiss brand is founded.

It is, of course, impossible to be exhaustive here. Illustrating the entire multiplicity of models from the course of the 125-year history of Breitling within these pages would be a near impossibility. Our goal, therefore, is to invite you to browse these pages and perhaps dream a little.

Heidelberg, August 2009
Henning Mützlitz

Ernest Schneider
successfully revived the
traditional brand in 1982.

The Breitling factory as seen in the year 1892.

The factory is located in Grenchen, Switzerland

ALL ABOARD

Eighteen eighty-four: the founding year of Breitling can be spied on each and every dial of the strikingly masculine wristwatches that this globally reputable brand hailing from Grenchen, Switzerland manufacturers. Its own history also plays an important role — and justifiably so.

At the age of 24, Léon Breitling, a man of German descent, opened his first workshop in St. Imier. It at first concentrated on the manufacture of industrial chronographs and precision counters. The young company grew quickly and in 1892, Breitling moved it not so very far away to La Chaux-de-Fonds in the Jura mountains — the cradle of Swiss watchmaking. One year before Breitling manufactured its first wristwatch in 1915, its founder passed away. His son Gaston succeeded him and began outfitting the first pilots of World War I with his instruments for the wrist.

Breitling established itself as a pioneer in the furthering development of timekeeping with the development of the first independent

chronograph button technology, and was therefore not randomly the first choice when airplane pilots were being outfitted with the best chronographs in the 1930s. These represented instruments necessary to precise navigation for the pilots. The Breitling chronographs had garnered a reputation for being robust and reliable companions — even withstanding the ever increasing acceleration forces that affect man and machine. Although Breitling — contrary to other manufacturers — did not produce its own movements, the company's reassembly specialists performed their work exemplarily and knew precisely how to accommodate the needs of pilots.

Gaston's son Willy took over the reins of the company in 1932, only five years after his father had done so; Gaston passed away in 1927.

Just two years later, the chronograph pioneers presented another groundbreaking innovation: with a second button on the case it was now possible to reset the timing and allow it to immediately begin running again. Thus, the accumulation of several short timings in series was made possible. Only then did this wrist chronograph receive its modern look, comprising a crown at 3 o'clock and buttons at 2 and 4 o'clock.

In the mid-1930s, the company began its lasting cooperation with worldwide aviation. In 1936, Breitling became the official supplier to the British Royal Air Force and in 1942 outfitter for the United States Army Air Forces. In the same year, the brand launched the Chronomat, the first chronograph to contain a circular slide rule, which offered a number of calculating operations to pilots. The successful model paved the way for the Navitimer, which was presented ten years later, and even

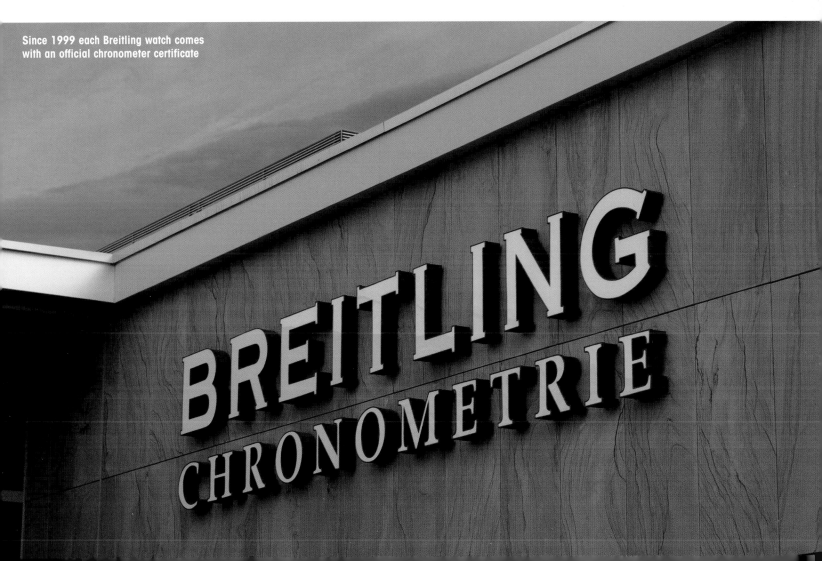

Since 1999 each Breitling watch comes
with an official chronometer certificate

offered extended calculation functions. A sort of mechanical navigation computer, the watch now made it possible for the pilot to perform navigational calculations in the cockpit simply by rotating the rotating ring under the crystal. The new chronograph quickly developed into a status symbol for pilots all over the world. The Navitimer, which conveys the essence of the Breitling brand as no other watch in the collection, has lost none of its status and remains the flagship for Breitling's entire range of models.

With the beginning of manned space travel, Breitling also left the earth's atmosphere: in 1962 a Cosmonaut model accompanied American astronaut Scott Carpenter into orbit in space capsule Aurora 7. In 1969, Breitling — in conjunction with Hamilton-Büren, Dubois Dépraz, and Heuer-Leonidas — introduced automatic winding for

Hollywood actor John Travolta is a prominent brand ambassador

chronographs. Now it was no longer necessary to inconveniently wind the mainspring of the watch by hand; the watch permanently wound itself using a rotor and the kinetic movement of its wearer. Caliber 11 proved its suitability for mass production and was ensuingly encased by Breitling about 300,000 times.

Despite this breakthrough in mechanical timekeeping, the successful brand Breitling also felt the effects of the Swiss industry's quartz crisis; the mechanics developed over the course of centuries now had to fight against the rapid expansion of cheap quartz, which led to a faltering in traditional production. On August 27, 1979, Willy Breitling announced that Breitling had to close its doors.

This closing was not meant to last, however. Engineer Ernest Schneider secured the rights to the Breitling and Navitimer names and registered his new company, Montres Breitling SA, in 1982 in Grenchen. At first,

The way to the future:
Breitling's own Caliber B01

it was specialized in manufacturing electronic watches. This era produced technical innovations such as the Aerospace model, an innovative multifunctional titanium chronograph. However, with the renaissance of mechanicals, the new-old company made an abrupt about-face, concentrating once again on traditional mechanical crafting. A re-issue of the Navitimer was presented in 1986, part of a phase of quality improvement and new developments that has continued to this day. In 1999, Breitling decided to offer all its watches only in certified chronometer quality — an expression of the highest possible quality.

Five years later, planning for the company's first in-house movement, the B01, had already begun. This represented a quantum leap in the development of the brand, and will ensure independence from external suppliers in the future. Breitling's watchmakers conceived a modern chronograph movement, which is not only simply constructed and thus suitable for mass production, but also fulfills the highest demands on precision. The new caliber was presented at the 2009 edition of Baselworld, the world's largest trade fair for watches and jewelry. By 2010 it is planned for use as the base for further developments.

Breitling's collection continues to consistently remain focused on instrument watches. Together with the brand's mature, inimitable design, its extremely high demands on quality, and its continued close relationship with aviation, it will build a fundament upon which the brand's continued success in the twenty-first century will be based.

The first chronograph with a button: Breitling's Mono-Poussoir model from 1923

HISTORICAL MODELS

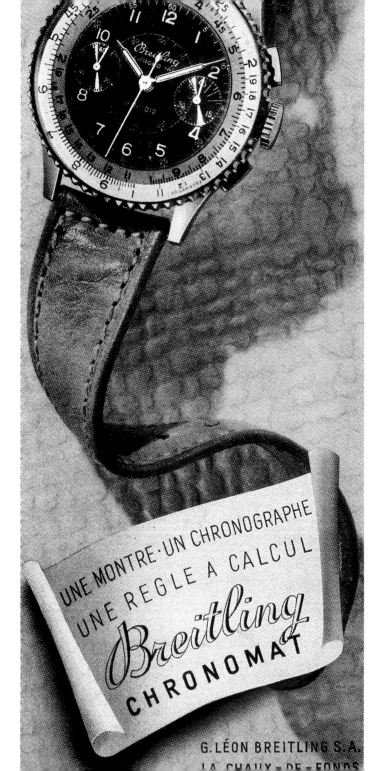

Over the course of its 125-year history, Breitling has brought a number of models and technical accomplishments to the market that have decisively advanced the development of the wristwatch in the twentieth century — both in terms of quartz and mechanics. Chronograph buttons, automatic winding for chronographs, and an ultra-precise super quartz movement are only some of the inventions proving the creativeness of the engineers and watchmakers employed by Breitling. Reliability and precision have always been the foremost considerations — two characteristics that stand for this brand like no others.

CHRONOGRAPH 1935

Reference number: 721
Movement: manually wound, Venus Caliber 170
Functions: hours, minutes, subsidiary seconds; chronograph
Case: chrome-plated stainless steel, ø 33 mm;
dial with tachymeter scale; push-down case back
Estimated value: $1530 (€ 1100,-)
(2009)

CHRONOGRAPH 1945

Movement: manually wound, Venus Caliber 170
Functions: hours, minutes, subsidiary seconds; chronograph
Case: stainless steel, ø 36 mm; screw-down case back
Estimated value: $980 (€ 700,-)
(2009)

ANTIMAGNETIC (1945)

Movement:	manually wound, Venus Caliber 188
Functions:	hours, minutes, subsidiary seconds; chronograph
Case:	stainless steel, ø 35 mm; dial with telemeter and tachymeter scales; push-down case back
Estimated value: (2009)	$1120 (€ 800,-)

CHRONOGRAPH (1979)

Movement:	manually wound, Venus Caliber 185
Functions:	hours, minutes, subsidiary seconds; split seconds chronograph; date, moon phase
Case:	yellow gold, ø 37 mm; push-down case back
Estimated value: (2009)	$11,180 (€ 8000,-)

CHRONOMAT (1960)

Reference number: 808
Movement: manually wound, Venus Caliber 175
Functions: hours, minutes, subsidiary seconds;
chronograph; slide rule function
Case: stainless steel, ø 37 mm; rotating bezel;
push-down case back
Estimated value: $4190 (€ 3000,-)
(2009)

CHRONOMAT (1963)

Reference number: 2110
Movement: automatic, Breitling Caliber 11
Functions: hours, minutes, chronograph; date
Case: stainless steel, ø 40 mm; rotating bezel with 60-minute
and 12-hour divisions; screw-down case back;
dial with tachymeter scale
Estimated value: $1260 (€ 900,-)
(2009)

CHRONOMAT (1973)

Reference number: 7808
Movement: manually wound, Valjoux Caliber 7740
Functions: hours, minutes, subsidiary seconds;
chronograph; date; slide rule function
Case: stainless steel, ø 41 mm; rotating bezel;
push-down case back; dial with tachymeter scale
Estimated value: $2935 (€ 2100,-)
(2009)

CHRONOMAT CHRONO-MATIC (1975)

Reference number: 8808.3
Movement: automatic, Breitling Caliber 12
Functions: hours, minutes, chronograph; date; slide rule function
Case: stainless steel, ø 41 mm; rotating bezel;
push-down case back; dial with tachymeter scale
Estimated value: $3075 (€ 2200,-)
(2009)

NAVITIMER (1969)

Reference number: 806
Movement: manually wound, Venus Caliber 178
Functions: hours, minutes, subsidiary seconds;
chronograph; slide rule function
Case: stainless steel, ø 40 mm; rotating bezel;
push-down case back; dial with tachymeter scale
Estimated value: $3075 (€ 2200,-)
(2009)

NAVITIMER (1973)

Reference number: 7806
Movement: manually wound, Valjoux Caliber 7740
Functions: hours, minutes, subsidiary seconds; date;
chronograph; slide rule function
Case: stainless steel, ø 40 mm; rotating bezel;
push-down case back; dial with tachymeter scale
Estimated value: $2795 (€ 2000,-)
(2009)

COSMONAUT (1975)

Reference number: 809.4
Movement: manually wound, Venus Caliber 178
Functions: hours, minutes, subsidiary seconds;
chronograph; slide rule function
Case: gold-plated stainless steel, ø 40 mm; rotating bezel;
push-down case back; dial with tachymeter scale
Estimated value: $3495 (€ 2500,-)
(2009)

COSMONAUT SUPER OCEAN (1975)

Reference number: 2105
Movement: automatic, Breitling Caliber 12
Functions: hours, minutes, chronograph; date
Case: stainless steel, ø 49 mm; rotating bezel; screw-down
case back; inner rotating ring with 60-minute divisions
Estimated value: $4475 (€ 1400,-)
(2009)

SUPER OCEAN REGATTA (1968)

Reference number: 7652
Movement: manually wound, Venus Caliber 178
Functions: hours, minutes, subsidiary seconds; chronograph; regatta timer
Case: stainless steel, ø 48 mm; rotating bezel; screw-down case back
Estimated value: $2235 (€ 1600,-)
(2009)

CHRONO-MATIC (1969)

Reference number: 2112
Movement: automatic, Breitling Caliber 11
Functions: hours, minutes, chronograph; date
Case: stainless steel, ø 39 mm; rotating bezel with 60-minute divisions; screw-down case back; dial with tachymeter scale
Estimated value: $840 (€ 600,-)
(2009)

CHRONOGRAPH TOP TIME (1969)

Reference number: 810
Movement: manually wound, Venus Caliber 178 TJ
Functions: hours, minutes, subsidiary seconds; chronograph
Case: stainless steel, ø 38 mm; push-down case back;
dial with tachymeter scale
Estimated value: $1675 (€ 1200,-)
(2009)

CHRONOGRAPH LONG PLAYING (1965)

Reference number: 7193.3
Movement: manually wound, Valjoux Caliber 7740
Functions: hours, minutes, subsidiary seconds; chronograph; date
Case: blackened stainless steel, 40 x 47 mm;
push-down case back
Estimated value: $840 (€ 600,-)
(2009)

CHRONOGRAPH FOOTBALL (1975)

Reference number: 2734.3
Movement: manually wound, Valjoux Caliber 7731
Functions: hours, minutes, subsidiary seconds; chronograph
Case: stainless steel, 41 x 47 mm; push-down case back;
     inner rotating ring with 60-minute divisions
Estimated value: $1260 (€ 900,-)
(2009)

CHRONOGRAPH SPRINT (1975)

Reference number: 1155
Movement: manually wound, Valjoux Caliber 7733
Functions: hours, minutes, subsidiary seconds; chronograph
Case: stainless steel, 40 x 43 mm; plastic bezel with
     tachymeter scale; push-down case back
Estimated value: $700 (€ 500,-)
(2009)

CALENDAR WATCH (1950)

Movement:	manually wound, Caliber 20
Functions:	hours, minutes, subsidiary seconds; date, weekday, month
Case:	yellow gold, ø 36 mm; push-down case back
Estimated value: **(2009)**	$1120 (€ 800,-)

MEN'S WATCH (1950)

Movement:	manually wound, Felsa Caliber 4010
Functions:	hours, minutes, sweep seconds
Case:	gold-plated stainless steel, ø 33 mm; push-down case back
Estimated value: **(2009)**	$210 (€ 150,-)

Breitling

De la qualité en série
Quality produced in series
Qualität in serien

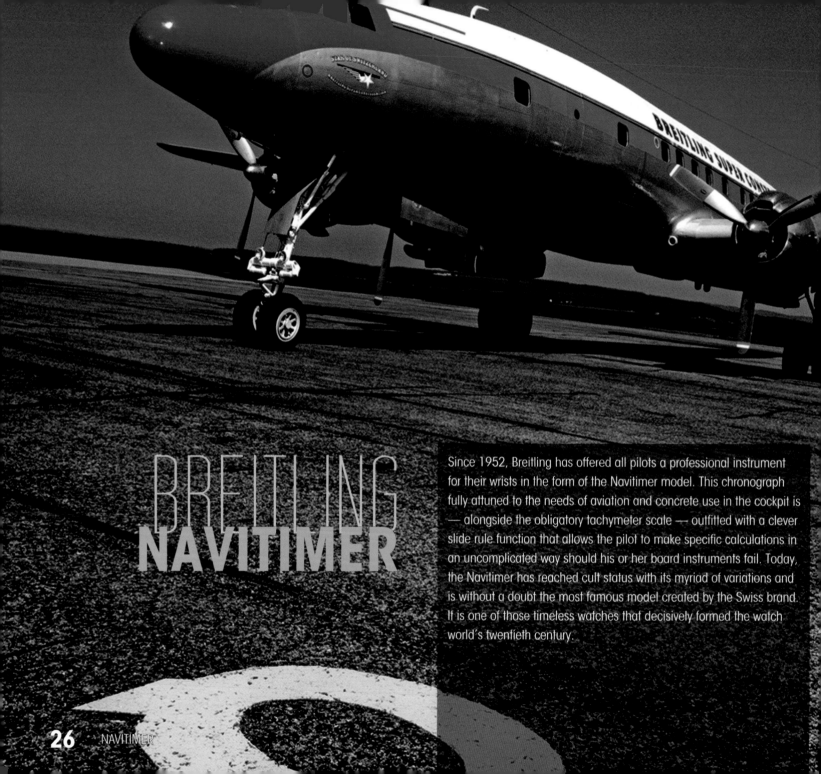

BREITLING
NAVITIMER

Since 1952, Breitling has offered all pilots a professional instrument for their wrists in the form of the Navitimer model. This chronograph fully attuned to the needs of aviation and concrete use in the cockpit is — alongside the obligatory tachymeter scale — outfitted with a clever slide rule function that allows the pilot to make specific calculations in an uncomplicated way should his or her board instruments fail. Today, the Navitimer has reached cult status with its myriad of variations and is without a doubt the most famous model created by the Swiss brand. It is one of those timeless watches that decisively formed the watch world's twentieth century.

NAVITIMER

Reference number: A23322-161

Movement: automatic, Breitling Caliber 23; officially certified chronometer (COSC)

Functions: hours, minutes, subsidiary seconds; chronograph; date

Case: stainless steel, ø 41.8 mm; unidirectionally rotating bezel with slide rule function and tachymeter scale; sapphire crystal

Price (2008): $5675 (€ 3890,-)

TECHNICAL DATA

NAVITIMER HERITAGE

Reference number: J35350-017
Movement: automatic, Breitling Caliber 35;
officially certified chronometer (COSC)
Functions: hours, minutes, subsidiary seconds; chronograph; date
Case: white gold, ø 43 mm; bidirectionally rotating bezel with
slide rule function and tachymeter scale; sapphire crystal
Price (2006): $26,290 (€ 22.200,-)

NAVITIMER HERITAGE

Reference number: J35340-015
Movement: automatic, Breitling Caliber 35;
officially certified chronometer (COSC)
Functions: hours, minutes, subsidiary seconds; chronograph; date
Case: white gold, ø 46 mm; bidirectionally rotating bezel with
slide rule function and tachymeter scale; sapphire crystal
Price (2002): $3490 (€ 3940,-)

NAVITIMER WORLD

Reference number: A24322-101
Movement: automatic, Breitling Caliber 24;
officially certified chronometer (COSC)
Functions: hours, minutes, subsidiary seconds; chronograph;
date; 24-hour display (second time zone)
Case: stainless steel, ø 46 mm; bidirectionally rotating bezel
with slide rule function and tachymeter scale
Price (2008): $6115 (€ 4190,-)

NAVITIMER MONTBRILLANT

Reference number: R41370
Movement: automatic, Breitling Caliber 41; officially certified chronometer (COSC)
Functions: hours, minutes, subsidiary seconds; chronograph; date
Case: rose gold, ø 38 mm; bidirectionally rotating bezel with tachymeter scale; sapphire crystal
Price (2009): $14,155 (€ 10.130,-)

NAVITIMER MONTBRILLANT

Reference number: A41330-101
Movement: automatic, Breitling Caliber 41; officially certified chronometer (COSC)
Functions: hours, minutes, subsidiary seconds; chronograph; date
Case: rose gold, ø 38 mm; bidirectionally rotating bezel with tachymeter scale; sapphire crystal
Price (2008): $5080 (€ 3480,-)

NAVITIMER MONTBRILLANT DATORA

Reference number: A21330-045
Movement: automatic, Breitling Caliber 21;
officially certified chronometer (COSC)
Functions: hours, minutes, subsidiary seconds; chronograph;
date, weekday, month; 24-hour display
Case: stainless steel, ø 43 mm; bidirectionally rotating
bezel with slide rule function and tachymeter scale;
sapphire crystal
Price (2008): $7470 (€ 5120,-)

NAVITIMER MONTBRILLANT DATORA

Reference number: A21330
Movement: automatic, Breitling Caliber 21;
officially certified chronometer (COSC)
Functions: hours, minutes, subsidiary seconds; chronograph; date,
weekday, month; 24-hour display (second time zone)
Case: stainless steel, ø 43 mm; bidirectionally rotating bezel
with integrated slide rule and tachymeter scale;
sapphire crystal
Price (2009): $5825 (€ 4710,-)

NAVITIMER MONTBRILLANT LÉGENDE

Reference number: C23340-046

Movement: automatic, Breitling Caliber 23; officially certified chronometer (COSC)

Functions: hours, minutes, subsidiary seconds; chronograph; date

Case: stainless steel, ø 47 mm; unidirectionally rotating rose gold bezel with slide rule function and tachymeter scale; sapphire crystal

Price (2007): $9540 (€ 7230,-)

OLD NAVITIMER

Reference number: A13322-161
Movement: automatic, Breitling Caliber 13;
                   officially certified chronometer (COSC)
Functions: hours, minutes, subsidiary seconds; chronograph; date
Case: stainless steel, ø 41.5 mm; bidirectionally rotating bezel
                   with slide rule function and tachymeter scale;
                   sapphire crystal
Price (2002): $2510 (€ 2830,-)

NAVITIMER 50TH ANNIVERSARY

Reference number: A41322-015
Movement: automatic, Breitling Caliber 41;
                   officially certified chronometer (COSC)
Functions: hours, minutes, subsidiary seconds; chronograph; date
Case: stainless steel, ø 41.8 mm; bidirectionally rotating
                   bezel with slide rule function and tachymeter scale;
                   sapphire crystal
Price (2002): $3590 (€ 4050,-)

NAVITIMER MONTBRILLANT 1903

Reference number: K35330-331
Movement: automatic, Breitling Caliber 35; officially certified chronometer (COSC)
Functions: hours, minutes, subsidiary seconds; chronograph; date
Case: red gold, ø 42 mm; bidirectionally rotating bezel with slide rule function and tachymeter scale; sapphire crystal
Price (2003): $11,065 (€ 10.550,-)

NAVITIMER MONTBRILLANT OLYMPUS

Reference number: A19350-3512
Movement: automatic, Breitling Caliber 19;
officially certified chronometer (COSC)
Functions: hours, minutes, subsidiary seconds; chronograph;
calendar (programmed for four years) with date,
weekday, month, moon phase
Case: stainless steel, ø 43 mm; bidirectionally rotating
bezel with slide rule function and tachymeter scale;
sapphire crystal
Price (2008): $8490 (€ 5820,-)

NAVITIMER MONTBRILLANT OLYMPUS

Reference number: A19340-011
Movement: automatic, Breitling Caliber 19;
officially certified chronometer (COSC)
Functions: hours, minutes, subsidiary seconds; chronograph;
calendar (programmed for four years) with date,
weekday, month, moon phase
Case: stainless steel, ø 43 mm; bidirectionally rotating
bezel with slide rule function and tachymeter scale;
sapphire crystal
Price (2004): $5700 (€ 4530,-)

NAVITIMER COSMONAUTE

Reference number: R22322-1612
Movement: automatic, Breitling Caliber 22;
officially certified chronometer (COSC)
Functions: hours, minutes, subsidiary seconds; chronograph; date
Case: red gold, ø 41.5 mm; bidirectionally rotating bezel
with slide rule function and tachymeter scale;
sapphire crystal; water-resistant to 3 atm
Price (2009): $15,550 (€ 11.130,-)

NAVITIMER COSMONAUTE

Reference number: R22322-1612
Movement: automatic, Breitling Caliber 22;
officially certified chronometer (COSC)
Functions: hours, minutes, subsidiary seconds; chronograph; date
Case: red gold, ø 41.5 mm; bidirectionally rotating bezel
with slide rule function and tachymeter scale;
sapphire crystal; water-resistant to 3 atm
Price (2009): $6275 (€ 4490,-)

TECHNICAL DATA

NAVITIMER CHRONO-MATIC 49

Reference number: A14360-011
Movement: automatic, Breitling Caliber 14; officially certified chronometer (COSC)
Functions: hours, minutes, subsidiary seconds; chronograph; date
Case: stainless steel, ø 49 mm; bidirectionally rotating rubber bezel with slide rule function and tachymeter scale; sapphire crystal
Price (2008): $6975 (€ 4780,-)

NAVITIMER CHRONO-MATIC 1461

Reference number: A19360
Movement: automatic, Breitling Caliber 19; officially certified chronometer (COSC)
Functions: hours, minutes, subsidiary seconds; chronograph; date, weekday, month
Case: stainless steel, ø 49 mm; bidirectionally rotating rubber-coated bezel with integrated slide rule; sapphire crystal
Price (2009): $10,150 (€ 7680,-)

BREITLING
AVENGER

According to Breitling's philosophy, the concept of its high-performance chronographs rests on four fundamental characteristics: robustness, functionality, precision, and aesthetics. The Avenger chronograph is a classic example: housed in a solid case characterized by well thought-out ergonomics, the Avenger also contains excellent legibility of all its displays. The quartz and automatic movements are all certified chronometers. And, lastly, the reserved, though catchy, design of the Avenger models fulfills the four self-defined prerequisites.

AVENGER SKYLAND

Reference number: A13380
Movement: automatic, Breitling Caliber 13; officially certified chronometer (COSC)
Functions: hours, minutes, subsidiary seconds; chronograph; date
Case: stainless steel, ø 45 mm; unidirectionally rotating bezel with 60-minute divisions; sapphire crystal; screw-in crown; water-resistant to 30 atm
Price (2009): $4600 (€ 3290,-)

CHRONO AVENGER M1

Reference number: E73360-2014
Movement: quartz, Breitling Caliber 73; officially certified chronometer (COSC)
Functions: hours, minutes, subsidiary seconds; chronograph; regatta configurations; date
Case: titanium, ø 44 mm; unidirectionally rotating bezel with 60-minute divisions; sapphire crystal; screw-in crown; water-resistant to 100 atm
Price (2006): $2840 (€ 2400,-)

AVENGER SKYLAND

Reference number: A13380
Movement: automatic, Breitling Caliber 13; officially certified chronometer (COSC)
Functions: hours, minutes, subsidiary seconds; chronograph; date
Case: stainless steel, ø 45 mm; unidirectionally rotating bezel with 60-minute divisions; sapphire crystal; screw-in crown; water-resistant to 30 atm
Price (2009): $4735 (€ 3390,-)

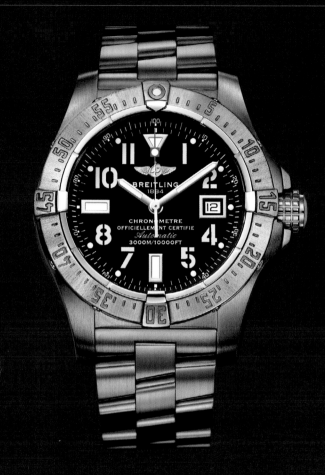

AVENGER SEAWOLF

Reference number: E17370-1014
Movement: automatic, Breitling Caliber 44; officially certified chronometer (COSC)
Functions: hours, minutes, sweep seconds; chronograph; date
Case: titanium, ø 44 mm; unidirectionally rotating bezel with 60-minute divisions; sapphire crystal; screw-in crown; water-resistant to 30 atm
Price (2006): $2425 (€ 2050,-)

AVENGER SEAWOLF

Reference number: A17330
Movement: automatic, Breitling Caliber 17; officially certified chronometer (COSC)
Functions: hours, minutes, sweep seconds; chronograph; date
Case: stainless steel, ø 44 mm; unidirectionally rotating bezel with 60-minute divisions; sapphire crystal; screw-in crown; water-resistant to 30 atm
Price (2009): $4095 (€ 2930,-)

AVENGER SEAWOLF CHRONO

Reference number: E73390

Movement: quartz, Breitling Caliber 73; officially certified chronometer (COSC)

Functions: hours, minutes, subsidiary seconds; chronograph; regatta configuration (10-minute countdown); date

Case: stainless steel, ø 45.4 mm; unidirectionally rotating bezel with 60-minute divisions; sapphire crystal; screw-in crown; water-resistant to 100 atm

Price (2009): $4360 (€ 3120,-)

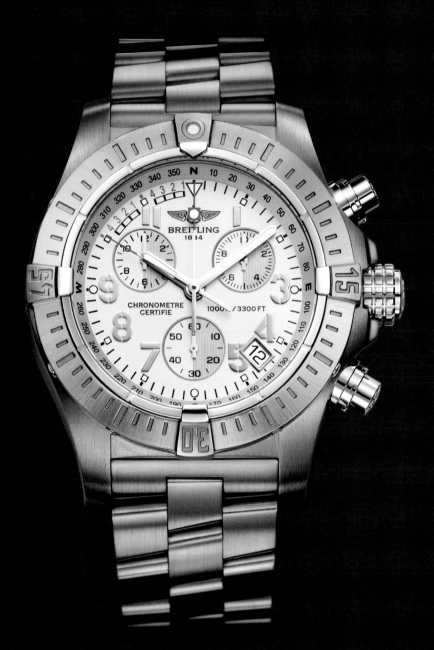

AVENGER SEAWOLF CHRONO

Reference number: E73390
Movement: quartz, Breitling Caliber 73;
officially certified chronometer (COSC)
Functions: hours, minutes, subsidiary seconds; chronograph;
regatta configuration (10-minute countdown); date
Case: stainless steel, ø 45.4 mm; unidirectionally rotating
bezel with 60-minute divisions; sapphire crystal;
screw-in crown; water-resistant to 100 atm
Price (2009): $4780 (€ 3420,-)

CHRONO AVENGER

Reference number: A13360-308
Movement: automatic, Breitling Caliber 13; officially certified chronometer (COSC)
Functions: hours, minutes, subsidiary seconds; chronograph; date
Case: titanium, ø 44 mm; unidirectionally rotating bezel with 60-minute divisions; sapphire crystal; screw-in crown; water-resistant to 30 atm
Price (2006): $3495 (€ 2950,-)

SUPER AVENGER

Reference number: A13370-168
Movement: automatic, Breitling Caliber 13;
                  officially certified chronometer (COSC)
Functions: hours, minutes, subsidiary seconds; chronograph; date
Case: stainless steel ø 48.4 mm; unidirectionally rotating
                  bezel with 60-minute divisions; sapphire crystal;
                  screw-in crown; water-resistant to 30 atm
Price (2008): $4640 (€ 3180,-)

SUPER AVENGER

Reference number: A13370-168
Movement: automatic, Breitling Caliber 13; officially certified chronometer (COSC)
Functions: hours, minutes, subsidiary seconds; chronograph; date
Case: stainless steel ø 48.4 mm; unidirectionally rotating bezel with 60-minute divisions; sapphire crystal; screw-in crown; water-resistant to 30 atm
Price (2009): $5100 (€ 3650,-)

BREITLING
CHRONOGRAPHS

Over the course of its 125-year history, Breitling has counted among the pioneers in the development of precise and performance-oriented chronographs. In 1915, the first wristwatch containing a stop function was presented, and in 1923 the first one-button chronograph debuted. The famous Chronomat and Navitimer models continued the tradition, and in 2009 Breitling presented its first in-house chronograph caliber, which is slated to go into large-scale production in the next few years. The Chronomat B01, outfitted with this new movement, is therefore in a way Breitling's own anniversary present to itself, guaranteeing the Swiss company a strong and independent position in the global watch industry in the future.

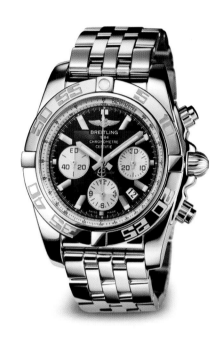

CHRONOMAT B01

Reference number: AB0110
Movement: automatic, Breitling Caliber B01;
officially certified chronometer (COSC)
Functions: hours, minutes, subsidiary seconds; chronograph; date
Case: stainless steel, ø 43.7 mm; unidirectionally rotating
bezel with 60-minute divisions; sapphire crystal;
screw-in crown; water-resistant to 50 atm
Price (2009): $8565 (€ 6130,-)

CHRONOMAT EVOLUTION

Reference number: B13356-086

Movement: automatic, Breitling Caliber 13;
officially certified chronometer (COSC)

Functions: hours, minutes, subsidiary seconds; chronograph; date

Case: stainless steel, ø 43.7 mm;
unidirectionally rotating bezel;
sapphire crystal; gold screw-in crown and buttons;
water-resistant to 30 atm

Price (2009): $9195 (€ 6580,-)

CHRONOMAT BLACKBIRD

Reference number: A13353-1014
Movement: automatic, Breitling Caliber 13; officially certified chronometer (COSC)
Functions: hours, minutes, subsidiary seconds; chronograph; date
Case: stainless steel, ø 44 mm; unidirectionally rotating bezel with 60-minute divisions; sapphire crystal; screw-in crown; water-resistant to 10 atm
Price (2002): $2235 (€ 2520,-)

CHRONOMAT

Reference number: A13352-257
Movement: automatic, Breitling Caliber 13; officially certified chronometer (COSC)
Functions: hours, minutes, subsidiary seconds; chronograph; date
Case: stainless steel, ø 40.5 mm; unidirectionally rotating bezel with 60-minute divisions; sapphire crystal; screw-in crown; water-resistant to 10 atm
Price (2002): $3015 (€ 3400,-)

TECHNICAL DATA

COLT GMT

Reference number: A32350-301
Movement: automatic, Breitling Caliber 32; officially certified chronometer (COSC)
Functions: hours, minutes, sweep seconds; date; 24-hour display (second time zone)
Case: stainless steel, ø 40.5 mm; unidirectionally rotating bezel with 60-minute divisions; sapphire crystal; screw-in crown; water-resistant to 50 atm
Price (2008): $2815 (€ 1930,-)

COLT GMT +

Reference number: A32370
Movement: automatic, Breitling Caliber 32; officially certified chronometer (COSC)
Functions: hours, minutes, sweep seconds; date; 24-hour display (second time zone)
Case: integrated stainless steel, ø 41.3 mm; unidirectionally rotating bezel with 60-minute divisions; sapphire crystal; screw-in crown; water-resistant to 50 atm
Price (2009): $3800 (€ 2720,-)

BLACKBIRD

Reference number: A44359
Movement: automatic, Breitling Caliber 44;
officially certified chronometer (COSC)
Functions: hours, minutes, subsidiary seconds;
chronograph; large date
Case: stainless steel, ø 43.7 mm; unidirectionally rotating
bezel with tachymeter scale; sapphire crystal;
screw-in crown and buttons; water-resistant to 30 atm
Price (2009): $7295 (€ 5220,-)

AIRWOLF RAVEN

Reference number: A78364
Movement: quartz, Breitling Caliber 78; officially certified chronometer (COSC)
Functions: hours, minutes, sweep seconds; chronograph; perpetual calendar with date, weekday, month, year; second time zone (world time); timer with alarm signal
Case: stainless steel, ø 43.5 mm; bidirectionally rotating stainless steel/rubber bezel with 360-degree divisions and compass directions; sapphire crystal; screw-in crown; water-resistant to 5 atm
Price (2009): $4765 (€ 3410,-)

AIRWOLF

Reference number: A78363-108
Movement: quartz, Breitling Caliber 78; officially certified chronometer (COSC)
Functions: hours, minutes, sweep seconds; chronograph; perpetual calendar with date, weekday, month, year; second time zone (world time); timer with alarm signal
Case: stainless steel, ø 43.5 mm; bidirectionally rotating bezel with 360-degree display; sapphire crystal; screw-in crown; water-resistant to 5 atm
Price (2008): $4730 (€ 3240,-)

SKYRACER RAVEN

Reference number: A27364-1018
Movement: automatic, Breitling Caliber 27;
officially certified chronometer (COSC)
Functions: hours, minutes, subsidiary seconds; chronograph; date
Case: stainless steel, ø 43.5 mm; bidirectionally rotating
rubber bezel with 60-minute divisions; sapphire crystal;
screw-in crown and buttons; water-resistant to 20 atm
Price (2009): $6385 (€ 4570,-)

TECHNICAL DATA

EMERGENCY MISSION

Reference number: A73322-018
Movement: quartz, Breitling Caliber 73; officially
certified chronometer (COSC)
Functions: hours, minutes, subsidiary seconds; chronograph; date;
micro transmitter outfitted with aviation
emergency frequency 121.5 MHz
Case: stainless steel, ø 45 mm; bidirectionally rotating rubber
bezel with 60-minute divisions; sapphire crystal;
screw-in crown; water-resistant to 10 atm
Price (2008): $6870 (€ 4710,-)

AEROSPACE

Reference number: K79362-1012
Movement: quartz, Breitling Caliber 79; officially certified chronometer (COSC)
Functions: hours, minutes (analog), chronograph; countdown; second time zone; alarm; timer with alarm signal
Case: yellow gold, ø 42 mm; unidirectionally rotating bezel with 60-minute divisions; sapphire crystal
Price (2009): $15,060 (€ 10.780,-)

B – ONE

Reference number: A68362-078
Movement: quartz, Breitling Caliber 68;
officially certified chronometer (COSC)
Functions: hours, minutes, sweep seconds (analog);
chronograph with digital display for intermediate and
addition timekeeping as well as a countdown function;
perpetual calendar with digital display for date, weekday,
month, year; second time zone, world time display;
timer with alarm signal
Case: stainless steel, ø 44.8 mm; rotating bezel with
60-minute divisions; water-resistant to 5 atm
Price (2003): $2410 (€ 2300,-)

B – ONE

Reference number: A78362-101U
Movement: quartz, Breitling Caliber 78;
officially certified chronometer (COSC)
Functions: hours, minutes, sweep seconds (analog);
chronograph with digital display for intermediate and
addition timekeeping as well as a countdown function;
perpetual calendar with digital display for date, weekday,
month, year; second time zone, world time display;
timer with alarm signal; additional 24-hour
analog time display on strap lug
Case: stainless steel, ø 43.2 mm; bidirectionally rotating
bezel with 60-minute divisions; sapphire crystal;
water-resistant to 5 atm
Price (2005): $4300 (€ 3180,-)

B – TWO

Reference number: A42362-108
Movement: automatic, Breitling Caliber 42;
officially certified chronometer (COSC)
Functions: hours, minutes, subsidiary seconds; chronograph; date
Case: stainless steel, ø 44.8 mm; bidirectionally rotating
bezel with 360-degree display; sapphire crystal;
water-resistant to 10 atm
Price (2003): $2950 (€ 2810,-)

B – TWO

Reference number: A42362-108
Movement: automatic, Breitling Caliber 42;
officially certified chronometer (COSC)
Functions: hours, minutes, subsidiary seconds; chronograph; date
Case: stainless steel, ø 44.8 mm; bidirectionally rotating
bezel with 360-degree display; sapphire crystal;
water-resistant to 10 atm
Price (2005): $3800 (€ 2810,-)

TECHNICAL DATA

HERCULES

Reference number: A39362-108
Movement: automatic, Breitling Caliber 39;
officially certified chronometer (COSC)
Functions: hours, minutes, subsidiary seconds;
chronograph; double minute counter
Case: stainless steel, ø 44.8 mm; bidirectionally rotating bezel
with 360-degree display; sapphire crystal; screw-in crown
Price (2002): $2525 (€ 2850,-)

CROSSWIND

Reference number: K13355-101
Movement: automatic, Breitling Caliber 13; officially certified chronometer (COSC)
Functions: hours, minutes, subsidiary seconds; chronograph; date
Case: yellow gold, ø 44 mm; unidirectionallyrotating bezel with 60-minute divisions; sapphire crystal; screw-in crown; water-resistant to 10 atm
Price (2003): $10,260 (€ 9780,-)

CROSSWIND SPECIAL

Reference number: K44355-251
Movement: automatic, Breitling Caliber 44; officially certified chronometer (COSC)
Functions: hours, minutes, subsidiary seconds; chronograph; large date
Case: stainless steel, ø 44 mm; unidirectionally rotating bezel with 60-minute divisions; sapphire crystal; screw-in crown; water-resistant to 10 atm
Price (2003): $11,640 (€ 11.100,-)

BREITLING
SUPEROCEAN

The top-performance diver's watch Superocean was created in the 1950s and was the timepiece of choice for special unit professional and military divers. With the growing popularity of diving, more and more hobby and leisure divers also decided to utilize the robust timekeeper. Today, you no longer need to be a diver to wear the Superocean, for the model has long become a classic lifestyle object and can be worn just as easily with a suit and a tie as with leisure clothing. And should the need arise, then the Superocean's monocoque case can boast water resistance to a full 4,900 feet (1,500 meters) — just in case.

TECHNICAL DATA

SUPEROCEAN

Reference number: A17360-078
Movement: automatic, Breitling Caliber 17;
officially certified chronometer (COSC)
Functions: hours, minutes, sweep seconds; date
Case: stainless steel, ø 42 mm; unidirectionally rotating
bezel with marking; sapphire crystal; screw-in crown;
water-resistant to 150 atm
Price (2009): $3185 (€ 2280,-)

SUPEROCEAN

Reference number: A17360-158
Movement: automatic, Breitling Caliber 17;
officially certified chronometer (COSC)
Functions: hours, minutes, sweep seconds; date
Case: stainless steel, ø 42 mm; unidirectionally
rotating bezel with marking; sapphire crystal;
screw-in crown; water-resistant to 150 atm
Price (2009): $3185 (€ 2280,-)

SUPEROCEAN STEELFISH

Reference number: A17390-308
Movement: automatic, Breitling Caliber 17; officially certified chronometer (COSC)
Functions: hours, minutes, sweep seconds; date
Case: stainless steel, ø 42 mm; unidirectionally rotating bezel with marking; sapphire crystal; screw-in crown; water-resistant to 200 atm
Price (2009): $3285 (€ 2350,-)

SUPEROCEAN STEELFISH

Reference number: A17390-171
Movement: automatic, Breitling Caliber 17; officially certified chronometer (COSC)
Functions: hours, minutes, sweep seconds; date
Case: stainless steel, ø 42 mm; unidirectionally rotating bezel with marking; sapphire crystal; screw-in crown; water-resistant to 200 atm
Price (2009): $2920 (€ 2090,-)

CHRONO SUPEROCEAN

Reference number: A13340-018
Movement: automatic, Breitling Caliber 13; officially certified chronometer (COSC)
Functions: hours, minutes, subsidiary seconds; chronograph; date, weekday
Case: stainless steel, ø 42 mm; unidirectionally rotating bezel with 60-minute divisions; sapphire crystal; screw-in crown and buttons; water-resistant to 50 atm
Price (2009): $4300 (€ 3080,-)

CHRONO SUPEROCEAN

Reference number: A13340-018
Movement: automatic, Breitling Caliber 13;
officially certified chronometer (COSC)
Functions: hours, minutes, subsidiary seconds;
chronograph; date, weekday
Case: stainless steel, ø 42 mm; unidirectionally rotating
bezel with 60-minute divisions; sapphire crystal;
screw-in crown and button; water-resistant to 50 atm
Price (2006): $3435 (€ 2900,-)

SUPEROCEAN HERITAGE 38

Reference number: A37320
Movement: automatic, Breitling Caliber 37; officially certified chronometer (COSC)
Functions: hours, minutes, subsidiary seconds; date
Case: stainless steel, ø 38 mm; unidirectionally rotating bezel with reference marker; sapphire crystal; screw-in crown; water-resistant to 20 atm
Price (2009): $3940 (€ 2820,-)

SUPEROCEAN HERITAGE 46

Reference number: A17320
Movement: automatic, Breitling Caliber 17;
officially certified chronometer (COSC)
Functions: hours, minutes, sweep seconds; date
Case: stainless steel, ø 38 mm; unidirectionally rotating
bezel with reference marker; sapphire crystal;
screw-in crown; water-resistant to 20 atm
Price (2009): $3630 (€ 2600,-)

SUPEROCEAN HERITAGE CHRONO

Reference number: A13320
Movement: automatic, Breitling Caliber 13; officially certified chronometer (COSC)
Functions: hours, minutes, seconds; chronograph; date
Case: stainless steel, ø 46 mm; unidirectionally rotating bezel with reference marker; sapphire crystal; screw-in crown; water-resistant to 20 atm
Price (2009): $5380 (€ 3850,-)

SUPEROCEAN HERITAGE CHRONOGRAPH LIMITED EDITION

Reference number: 23320-018
Movement: automatic, Breitling Caliber 23; officially certified chronometer (COSC)
Functions: hours, minutes, subsidiary seconds; chronograph; date
Case: stainless steel, ø 46 mm; unidirectionally rotating bezel with reference marker; sapphire crystal; screw-in crown; water-resistant to 20 atm
Price (2009): $5575 (€ 3990,-)

BREITLING
BREITLING FOR BENTLEY

A few years ago, Breitling rediscovered the street, and together with Bentley — one of the most prestigious automobile manufacturers in the world — has put together an exclusive and exceptional collection fittingly called Breitling for Bentley. This line of luxurious chronographs contains many style elements inspired by the English brand's noble cars: the dials and cases are offered in colors and woods typical of Bentley, some case backs are styled like light alloy wheel rims, and the finishing is reminiscent of the interior of the luxury limousines. The collection's most magnificent piece is the Mulliner Tourbillon, an homage to the famous British car body manufacturer.

BENTLEY MULLINER TOURBILLON

Reference number: K18841-0412
Movement: automatic, Breitling Caliber 18B; one-minute tourbillon; officially certified chronometer (COSC)
Functions: hours, minutes, subsidiary seconds; chronograph; date
Case: yellow gold, ø 48.7 mm; bidirectionally rotating bezel with slide rule function; sapphire crystal; transparent case back with precious wood decor; screw-in crown; water-resistant to 10 atm
Price (2006): $151,950 (€ 128.320,-)

BENTLEY MOTORS

Reference number: K25362
Movement: automatic, Breitling Caliber 25B; officially certified chronometer (COSC)
Functions: hours, minutes, subsidiary seconds; chronograph; date
Case: yellow gold, ø 48.7 mm; rotating bezel with slide rule function; sapphire crystal; screw-in crown; water-resistant to 10 atm
Price (2009): $48,540 (€ 34.740,-)

BENTLEY MOTORS T

Reference number: A25363-105
Movement: automatic, Breitling Caliber 25B; officially certified chronometer (COSC)
Functions: hours, minutes, subsidiary seconds; chronograph; date
Case: stainless steel, ø 48.7 mm; rotating bezel with slide rule function; sapphire crystal; screw-in crown; water-resistant to 10 atm
Price (2006): $7745 (€ 6540,-)

BENTLEY MULLINER PERPETUAL

Reference number: H29362-1012
Movement: automatic, Breitling Caliber 29B;
officially certified chronometer (COSC)
Functions: hours, minutes, subsidiary seconds; chronograph;
perpetual calendar with date, weekday,
month, moon phase, leap year
Case: rose gold, ø 48.7 mm; unidirectionally rotating
bezel with slide rule function; sapphire crystal;
screw-in crown; water-resistant to 5 atm
Price (2006): $46,600 (€ 39.360,-)

BENTLEY MARK VI

Reference number: P26362-0412
Movement: automatic, Breitling Caliber 26B;
officially certified chronometer (COSC)
Functions: hours, minutes, subsidiary seconds; chronograph; date
Case: stainless steel, ø 42 mm; platinum bezel;
sapphire crystal; screw-in crown; water-resistant to 5 atm
Price (2008): $13,280 (€ 9100,-)

BENTLEY GT

Reference number: A13362-025
Movement: automatic, Breitling Caliber 13B;
officially certified chronometer (COSC)
Functions: hours, minutes, subsidiary seconds; chronograph; date
Case: stainless steel, ø 44.8 mm; bidirectionally rotating
bezel with slide rule function; sapphire crystal;
screw-in crown; water-resistant to 5 atm
Price (2007): $7695 (€ 5830,-)

BENTLEY 6.75 SPEED

Reference number: A44364

Movement: automatic, Breitling Caliber 44B; officially certified chronometer (COSC)

Functions: hours, minutes, subsidiary seconds; chronograph; large date

Case: stainless steel, ø 48.7 mm; bidirectionally rotating bezel with slide rule function; sapphire crystal; screw-in crown; water-resistant to 5 atm

Price (2009): $8500 (€ 6090,-)

BENTLEY MOTORS SPEED

Reference number: A25364
Movement: automatic, Breitling Caliber 25B; officially certified chronometer (COSC)
Functions: hours, minutes, subsidiary seconds; chronograph; date
Case: stainless steel, ø 48.7 mm; sapphire crystal; screw-in crown; water-resistant to 5 atm
Price (2009): $8450 (€ 6050,-)

BENTLEY GMT

Reference number: A47362-0419
Movement: automatic, Breitling Caliber 47B;
officially certified chronometer (COSC)
Functions: hours, minutes, subsidiary seconds; chronograph;
date; 24-hour display (second time zone)
Case: stainless steel, ø 49 mm; bidirectionally rotating ring with
reference cities underneath the crystal; sapphire crystal;
screw-in crown; water-resistant to 5 atm
Price (2009): $9990 (€ 7150,-)

BENTLEY GMT

Reference number: A47362-0419
Movement: automatic, Breitling Caliber 47B;
officially certified chronometer (COSC)
Functions: hours, minutes, subsidiary seconds; chronograph;
date; 24-hour display (second time zone)
Case: stainless steel, ø 49 mm; bidirectionally rotating ring with
reference cities underneath the crystal; sapphire crystal;
screw-in crown; water-resistant to 5 atm
Price (2008): $10,050 (€ 6890,-)

THE FLYING B

Reference number: A28362-0612
Movement: automatic, Breitling Caliber 28B; officially certified chronometer (COSC)
Functions: hours (jump), minutes, subsidiary seconds
Case: stainless steel, 38.5 x 57.3 mm; sapphire crystal; transparent case back; screw-in crown; water-resistant to 5 atm
Price (2007): $14,425 (€ 10.930,-)

TECHNICAL DATA

FLYING B CHRONOGRAPH

Reference number: A44365-1012
Movement: automatic, Breitling Caliber 44B;
officially certified chronometer (COSC)
Functions: hours, minutes, subsidiary seconds;
chronograph; large date
Case: stainless steel, 44.38 x 58.23 mm; sapphire crystal;
screw-in crown; water-resistant to 5 atm
Price (2008): $15,950 (€ 10.930,-)

TECHNICAL DATA

NAVITIMER BENTLEY MOTORS

Reference number: A25362-105

Movement: automatic, Breitling Caliber 25B; officially certified chronometer (COSC)

Functions: hours, minutes, subsidiary seconds; chronograph with sweep 30-second hand; date

Case: stainless steel, ø 48.7 mm; bidirectionally rotating bezel with slide rule function and tachymeter scale; sapphire crystal; screw-in crown; water-resistant to 10 atm

Price (2003): $5840 (€ 5570,-)

BREITLING
COCKPIT

The exceptionally masculine Cockpit model, with its striking case design, was created by Breitling to handle situations where chronograph functions are really not necessary. Despite its name, the Cockpit is less targeted to the pilot's preferred place of work than to regular daily use. Robustness and sportiness as well as

excellent legibility — created in great part by the conspicuously positioned large date — make the Cockpit a reliable companion in every situation. Breitling now also offers this model outfitted with a chronograph — a matter of course for the traditionally minded brand at home in Grenchen, actually.

COCKPIT

Reference number: C49350-046

Movement: automatic, Breitling Caliber 49; officially certified chronometer (COSC)

Functions: hours, minutes, sweep seconds; large date

Case: stainless steel, ø 41 mm; unidirectionally rotating rose gold bezel with 60-minute divisions; sapphire crystal; screw-in crown; water-resistant to 30 atm

Price (2009): $6260 (€ 4480,-)

COCKPIT

Reference number: C49350-105
Movement: automatic, Breitling Caliber 49; officially certified chronometer (COSC)
Functions: hours, minutes, sweep seconds; large date
Case: stainless steel, ø 41 mm; unidirectionally rotating bezel with 60-minute divisions; sapphire crystal; screw-in crown; water-resistant to 30 atm
Price (2009): $4950 (€ 3540,-)

COCKPIT

Reference number: C49350-175
Movement: automatic, Breitling Caliber 49; officially certified chronometer (COSC)
Functions: hours, minutes, sweep seconds; large date
Case: stainless steel, ø 41 mm; unidirectionally rotating bezel with 60-minute divisions; sapphire crystal; screw-in crown; water-resistant to 30 atm
Price (2005): $4600 (€ 3400,-)

COCKPIT

Reference number: 49360
Movement: automatic, Breitling Caliber 49; officially certified chronometer (COSC)
Functions: hours, minutes, sweep seconds; large date
Case: rose gold, ø 41 mm; unidirectionally rotating bezel with 60-minute divisions; sapphire crystal; screw-in crown; water-resistant to 10 atm
Price: (2009): $5715 (€ 4090,-)

COCKPIT

Reference number: C49350

Movement: automatic, Breitling Caliber 49;
officially certified chronometer (COSC)

Functions: hours, minutes, sweep seconds; large date

Case: stainless steel, ø 41 mm; unidirectionally rotating
bezel with 60-minute divisions; sapphire crystal;
rose gold; screw-in crown; water-resistant to 30 atm

Price (2009): $6260 (€ 4480,-)

COCKPIT

Reference number: B49350-076
Movement: automatic, Breitling Caliber 49; officially certified chronometer (COSC)
Functions: hours, minutes, sweep seconds; large date
Case: stainless steel, ø 41 mm; unidirectionally rotating rose gold decorated bezel with 60-minute divisions; sapphire crystal; rose gold; screw-in crown; water-resistant to 30 atm
Price (2009): $7530 (€ 5390,-)

CHRONO COCKPIT

Reference number: B13358
Movement: automatic, Breitling Caliber 13;
officially certified chronometer (COSC)
Functions: hours, minutes, subsidiary seconds; chronograph; date
Case: stainless steel, ø 39 mm; unidirectionally rotating
stainless steel/gold bezel with 60-minute divisions;
screw-in crown; water-resistant to 10 atm
Price (2004): $3540 (€ 2810,-)

CHRONO COCKPIT

Reference number: 13357-011
Movement: automatic, Breitling Caliber 13;
officially certified chronometer (COSC)
Functions: hours, minutes, subsidiary seconds; chronograph; date
Case: stainless steel, ø 39 mm; unidirectionally rotating
bezel with 60-minute divisions; screw-in crown;
water-resistant to 10 atm
Price (2009): $8370 (€ 5990,-)

COCKPIT **93**

CHRONO COCKPIT

Reference number: B13358
Movement: automatic, Breitling Caliber 13;
officially certified chronometer (COSC)
Functions: hours, minutes, subsidiary seconds; chronograph; date
Case: stainless steel, ø 39 mm; unidirectionally rotating
bezel with 60-minute divisions; sapphire crystal;
screw-in crown; water-resistant to 10 atm
Price (2009): $8370 (€ 5990,-)

CHRONO COCKPIT

Reference number: B13358-237
Movement: automatic, Breitling Caliber 13;
officially certified chronometer (COSC)
Functions: hours, minutes, subsidiary seconds; chronograph; date
Case: rose gold, ø 39 mm; unidirectionally rotating stainless
steel/gold bezel with 60-minute divisions; sapphire
crystal; screw-in crown; water-resistant to 10 atm
Price (2009): $19,100 (€ 13.670,-)

BREITLING®